Vegetable Aroids

Growing Practices and Nutritional Information

Roby Jose Ciju

Table of Contents

Vegetable Aroids: An Introduction

Aroids are a group of useful plants belonging to the plant family Araceae. Though most of these aroid plants are used as ornamentals, some are used as nutritious vegetables. The most popular examples of ornamental aroids are aglonemas, money plants or pothos, monsteras and philodendrons. However, there are some other aroid plants which are highly useful as food plants. Some popular examples are taro or elephant's ear, yams, Amorphophallus or elephant foot yam, tannia or Xanthosoma spp., Lasia spp. or Kohila and Cyrtosperma spp. These aroids are mainly used as vegetables and hence the term 'vegetable aroids' are commonly used.

All these vegetables are easy to grow provided that they are growing in moist soils in moderate climatic conditions. Too much hot and too much cold condition is detrimental for the crop growth. These plants do not tolerate drought. All these crops grow well at an optimum temperature range of 20 -25 degree Celsius and in slightly acidic soils with pH ranging from 6.0 to 7.0.

Propagation is by corms or cormels. Sometimes setts or suckers are also used. Always go for healthy planting materials which are free of diseases and insects-pests. A detailed account of growing practices for these aroid vegetables is given below:

Figure 1: Taro tubers

Taro or Colocasia Tubers

Taro or colocasia is a popular tropical tuber crop grown for its edible root tubers and succulent, tender shoots and young leaves. Colocasia tubers are used as a staple food and as a vegetable in many countries. Colocasia leaves are used as a tropical leafy vegetable. There are many varieties of colocasia that are cultivated for edible purposes. Tubers of some varieties of colocasia may contain considerable amounts of an acrid compound called calcium oxalate which can be destroyed by proper cooking.

Figure 2: Colocasia plant

Origin and Taxonomy: Colocasia is believed to be

originated in the tropics of the old world comprising of the regions of southeastern Asia and the Indian subcontinent. Common names of colocasia are elephant-ear, taro, and coco yam. It belongs to the family Araceae, genus *Colocasia* and species *esculenta*. A detailed taxonomic classification of the plant is given below:

Kingdom:	Plantae
Order:	Alismatales
Family:	Araceae
Genus:	Colocasia
Species	C. esculenta

Plant Description: Colocasia plant is a dwarf-growing herbaceous perennial plant that arises from a large underground corm. Corm is a modified underground stem. Leaves are large and look like a large shield and sometimes reach up to 1-1.5 meter in length. Leaves are shaped like an arrowhead, with one point upwards and other two pointed lobes extending downwards. Its inflorescence is a spadix, a specialized spike of small, tiny flowers that are closely arranged round a fleshy axis and typically enclosed in a spathe, characteristic of all the arums (members of Family Araceae).

Figure 3: Taro Flower

Growing Colocasia/Taro: A detailed account of various growing practices for taro plant is given below:

Climate Requirements: Colocasia is a tropical tuber crop and can be successfully grown in warm, tropical climates. Two crops of colocasia are possible in tropical regions: Summer Crop (Dryland crop) and Rainy Season Crop (wetland crop).

Soil Requirements: Just like other tuber crops, colocasia needs well-drained, fertile, loose sandy loam soils for its healthy growth. Precisely speaking, for a Dryland crop, well-drained, well-aerated clay loam soils with a pH range of 5.5 to 6.5 are preferred. Water-logging must be avoided by all means. Wetland crop can be planted in irrigated fields.

Field Preparation: Prepare the field by continuous ploughing followed by furrow or ridge making.

Propagation and Planting: Colocasia is propagated through setts or cormels and tuber/corm cuttings. Sprouted tuber cuttings are planted in the main field for raising a colocasia crop. Summer crop is sown in February-March in tropics while the rainy season crop in planted in June-July. Colocasia tubers need plenty of moisture in the soil for vigorous sprouting and leaf production.

Spacing: Generally speaking, planting density of a rainy season crop is higher than that of a Dryland or summer crop. For a wetland crop or rainy season crop, a spacing of 45 cm x 45 cm may be adopted (i.e. planting density of 50,000 plants per hectare). Whereas for a summer crop, a spacing of 50 cm between plants and 1 meter between rows (20 000 plants per ha) may be adopted

Watering and Intercultural Operations: Colocasia plants need moist soil throughout its life cycle. Once sprouted-tuber cuttings are planted, care should be taken not to dry the soil. If planted tuber cuttings are not germinated properly, one light irrigation may help quick germination process. In an established crop, intercultural operations such as hoeing, earthing up soil, weeding etc may be carried out regularly. One or two light earthing up operations are beneficial for a colocasia crop.

Fertilizers and Manures: Colocasia crop may be raised in a well-prepared field where plenty of organic manures and compost may be mixed thoroughly with the top soil to replenish the soil fertility. Standard NPK fertilizers may be applied as per crop requirements. As a rule, apply fertilizers and manures as split doses; first dose at planting, and second dose 90-120 days after planting while corm development is happening.

Weed Management: Various weed management practices such as hand weeding, mechanical weeding; application of herbicides and mulching may be used for weed control in the fields.

Disease Management: Colocasia is a hardy plant and it withstands many insect-pest attacks. Generally there are no major diseases that attack colocasia plants in a serious way. However in some regions, colocasia is found to be susceptible to a disease called '*Colocasia Blight*' or taro leaf blight. It is a fungal infection. Colocasia blight attacks young leaves and petioles first and in later stages, tubers are also affected. Spraying Bordeaux mixture may control the disease very effectively. Other diseases may include viral infections and root and corm rots caused by fungal infections.

Insect Pest Management: Major insect-pests attacking the colocasia plants include delphacid plant hopper, Tarophagus spp.; Taro hawkmoth caterpillars, Hippotion celerio; cluster caterpillars, Papuana beetles; and podoptera litura.

Harvesting and Yield: Colocasia crop is ready for harvesting within four to five months after planting. In other words, colocasia crop matures in about 130 to 140 days after planting. Maturity is indicated by reduction in the height of the mother plant and the yellowing of leaves. Estimated yield is about 15 tons of tubers per hectare.

Note: Conversion Table

1 ton	1000 kilograms or 10 quintals
1 hectare/ha	10000 sq. m. (square meter)

Food Uses and Nutrition of Taro Tubers: Colocasia tubers can be cooked in many ways such as by boiling, baking, steaming, deep frying etc.

Taro chips: Sliced colocasia tubers may be deep-fried to make delicious colocasia chips.

Figure 4: Taro or Colocasia chips

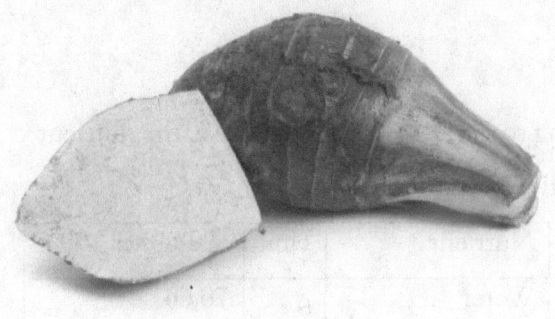

Figure 5: Taro Tubers

Taro cake:

Figure 6: Taro cake

According to USDA nutrient database, the nutrient composition of taro tubers is as follows:

Nutrient	Unit	Value per 100 g
Water	g	70.64
Energy	kcal	112
Protein	g	1.5
Total lipid (fat)	g	0.2
Carbohydrate	g	26.46
Fiber	g	4.1
Sugars, total	g	0.4

Calcium, Ca	mg	43
Iron, Fe	mg	0.55
Magnesium, Mg	mg	33
Phosphorus, P	mg	84
Potassium, K	mg	591
Sodium, Na	mg	11
Zinc, Zn	mg	0.23
Vitamin C	mg	4.5
Thiamin	mg	0.095
Riboflavin	mg	0.025
Niacin	mg	0.6
Vitamin B-6	mg	0.283
Folate, DFE	µg	22
Vitamin B-12	µg	0
Vitamin A	IU	76
Vitamin E	mg	2.38
Vitamin K	µg	1

Colocasia (Taro) Leaves

Young, tender leaves and shoots of colocasia plants may be used as a leafy vegetable. Colocasia/taro leaves are cooked like other green leafy vegetables such as spinach and mustard greens. Taro leaves are cooked in various forms; it may be baked, boiled to make soups and/or may be cooked with coconut, pulses, fish or meat to prepare various dishes. In many tropical countries, various curry (*dal*) preparations are made from a mixture of chopped taro leaves and various pulses.

Taro leaves are considered as a highly nutritious vegetable which is low in fatty acids and cholesterol and high in calcium, potassium and other essential minerals. It is a rich source of Vitamin E and Vitamin K.

Figure 7: Taro Leaves

Nutrition in Raw Taro Leaves: Taro leaves are an excellent source of vitamin A. According to U.S. FDA (Food and Drug Administration), daily value (DV) of Vitamin A is 5000 IU (international unit). 100 g of edible portion of taro leaves contains **4825** IU of Vitamin A.

Note: DV means daily amount of nutrient recommended for an adult

Vitamin A is essential for eye health and also for strengthening body's natural immune system. Vitamin A is also essential for tissue building, and for the formation of RBCs (red blood cells), skin and bones. Deficiency of Vitamin A results in night blindness, and drying of skin and eyes

Taro leaves are excellent source of vitamin K also. According to U.S. FDA (Food and Drug Administration), daily value

(DV) of Vitamin K is 80mcg (micrograms). 100g of edible portion of taro leaves contain **108.6** mcg of Vitamin K. Vitamin K is essential for the formation of strong bones and blood clotting. A detailed account of nutrients present in raw colocasia leaves is given below:

Nutrient	Unit	Value per100g
Water	g	85.66
Energy	kcal	42
Protein	g	4.98
Total lipid (fat)	g	0.74
Carbohydrate	g	6.7
Fiber, total dietary	g	3.7
Sugars, total	g	3.01
Calcium, Ca	mg	107
Iron, Fe	mg	2.25
Magnesium, Mg	mg	45
Phosphorus, P	mg	60
Potassium, K	mg	648
Sodium, Na	mg	3
Zinc, Zn	mg	0.41
Vitamin C	mg	52
Thiamin	mg	0.209
Riboflavin	mg	0.456

Niacin	mg	1.513
Vitamin B-6	mg	0.146
Folate, DFE	µg	126
Vitamin A	IU	4825
Vitamin E	mg	2.02
Vitamin K	µg	108.6

Similarly, nutrition in cooked/steamed taro leaves (as per USDA nutrient database) is as given below:

Nutrient	Unit	Value per100g	
		without salt	with salt
Water	g	92.15	92.15
Energy	kcal	24	24
Protein	g	2.72	2.72
Total lipid (fat)	g	0.41	0.41
Carbohydrate	g	4.02	3.89
Fiber, total dietary	g	2	2
Calcium, Ca	mg	86	86
Iron, Fe	mg	1.18	1.18
Magnesium, Mg	mg	20	20
Phosphorus, P	mg	27	27
Potassium, K	mg	460	460
Sodium, Na	mg	2	238

Zinc, Zn	mg	0.21	0.21
Vitamin C	mg	35.5	35.5
Thiamin	mg	0.139	0.139
Riboflavin	mg	0.38	0.38
Niacin	mg	1.267	1.267
Vitamin B-6	mg	0.072	0.072
Folate, DFE	µg	48	48
Vitamin A	IU	4238	4238

Nutrients in Raw Taro Shoots: Young shoots of taro plants are also used as leafy vegetables. As per USDA, the nutrient composition of young shoots of taro plants is as given below:

Nutrient	Unit	Value per 100 g
Proximates		
Water	g	95.82
Energy	kcal	11
Protein	g	0.92
Total lipid (fat)	g	0.09
Carbohydrate, by difference	g	2.32
Minerals		
Calcium, Ca	mg	12
Iron, Fe	mg	0.6

Magnesium, Mg	mg	8
Phosphorus, P	mg	28
Potassium, K	mg	332
Sodium, Na	mg	1
Zinc, Zn	mg	0.51
Vitamins		
Vitamin C, total ascorbic acid	mg	21
Thiamin	mg	0.04
Riboflavin	mg	0.05
Niacin	mg	0.8
Vitamin B-6	mg	0.111
Folate, DFE	µg	3
Vitamin B-12	µg	0
Vitamin A, RAE	µg	3
Vitamin A, IU	IU	50
Vitamin D (D2 + D3)	µg	0
Vitamin D	IU	0
Lipids		
Fatty acids, total saturated	g	0.018
Fatty acids, total monounsaturated	g	0.007
Fatty acids, total polyunsaturated	g	0.037
Fatty acids, total trans	g	0

Yams or Dioscorea Tubers

Yams are a group of monocot plants, precisely speaking, a group of perennial herbaceous vines grown for their starch-rich, edible tubers. They belong to the family Dioscoreaceae and genus *Dioscorea*. Yams are widely grown in both temperate and tropical parts of the world and considered as an important food crop in the tropical areas of America, Asia and Africa. Yams are a rich source of carbohydrates. There are white yams, purple yams and yellow yams, depending on the colour of the tuber flesh.

Figure 8: A Young Yam Plant

Dioscorea Species: The genus *Dioscorea* contains about 250 species of plants of which all species do not produce edible tubers. Some are weeds or invasive plants while others are cultivated for edible tubers. Among the species cultivated are *Dioscorea alata, Dioscorea esculenta,* and *Dioscorea caryenensis.*

In many parts of tropical Asian countries, purple yam is the most popular variety of Dioscorea tubers. Scientific name of purple yam is *Dioscorea alata.* It is also known as greater yam. The tuber flesh of greater yam is purple and therefore the name, 'purple yam.' These plants produce large edible tubers, some of which weigh up to 30 kg. Large tubers are used for propagation purposes while small and medium tubers are used for food purposes.

D. esculenta is the most commonly grown Dioscorea species in the Indian subcontinent. Tubers of these plants are white coloured. *Dioscorea caryenensis* is known as 'yellow yam' because tubers of these plants are yellow in colour.

Other Species of Yams: *D. batatas* (Chinese yam) and *D. bulbifera,* (air potato) are also popular among yam-growers. These yams are grown for their edible aerial tubers.

Figure 9: Yams

Origin and Distribution: Yams are believed to be a native of Africa, Asia and America. In some parts of these regions, yams grow naturally as invasive plants. A detailed taxonomic classification of purple yam is as given below:

Kingdom:	Plantae
Family:	Dioscoreaceae
Genus:	Dioscorea
Species	D. alata

Food Uses of Yams: Yams are used as vegetables and also as a staple food. Yam tubers can be consumed after baking, boiling, steaming, stir frying, microwave cooking, and roasting. Yams contain some acrid/pungent compounds which are eliminated while cooking by boiling or steaming. As per USDA nutrient database, the nutrient composition of raw yams is as follows:

Nutrient	Unit	Value per 100 g
Proximates		
Water	g	69.6
Energy	kcal	118
Protein	g	1.53
Total lipid (fat)	g	0.17
Carbohydrate, by difference	g	27.88
Fiber, total dietary	g	4.1
Sugars, total	g	0.5
Minerals		
Calcium, Ca	mg	17
Iron, Fe	mg	0.54
Magnesium, Mg	mg	21
Phosphorus, P	mg	55
Potassium, K	mg	816
Sodium, Na	mg	9
Zinc, Zn	mg	0.24
Vitamins		
Vitamin C, total ascorbic acid	mg	17.1

Thiamin	mg	0.112
Riboflavin	mg	0.032
Niacin	mg	0.552
Vitamin B-6	mg	0.293
Folate, DFE	µg	23
Vitamin B-12	µg	0
Vitamin A, RAE	µg	7
Vitamin A, IU	IU	138
Vitamin E (alpha-tocopherol)	mg	0.35
Vitamin D (D2 + D3)	µg	0
Vitamin D	IU	0
Vitamin K (phylloquinone)	µg	2.3
Lipids		
Fatty acids, total saturated	g	0.037
Fatty acids, total monounsaturated	g	0.006
Fatty acids, total polyunsaturated	g	0.076
Fatty acids, total trans	g	0
Cholesterol	mg	0
Amino Acids		
Other		
Caffeine	mg	0

Yams are often confused with sweet potatoes. But true yams, which differ in tuber sizes considerably (from the size of a small potato to the size of over 60-65 Kg tuber), are entirely different from sweet potatoes.

Growing Requirements: A detailed account of growing practices for yams is as given below:

Climate: Yams are tropical plants and hence warm, tropical climate is best for its healthy growth. Plenty of sunlight is needed for these plants.

Soil: Deep, rich, friable, well-drained and loose sandy loam soils are the most ideal for growing yams.

Propagation: Propagation is by division of tubers. These tuber cuttings are planted in a well-prepared nursery beds to allow sprouting. Sprouted tuber cuttings are then transplanted in the main field in pits/holes of appropriate size (at least 5 inches deep and 3 inches wide). While planting care is taken to place the sprouted leaves above the ground level.

Watering: Soon after planting, water the plants. Watering should be done regularly until plants get established in the field. Thereafter, need-based watering is done.

Manure Requirements: Just like other tuber crops, yams also need nitrogen for its vegetative growth and plenty of phosphorous for tuber development. Large quantities of farm yard manure or compost may be added to the top soil to replenish the soil fertility at the time of field preparation. Fertilizers for yams should be high in phosphorous.

Spacing: Under right growing conditions, yam plants grow vigorously and need sufficient space for tuber production. Hence it is recommended that at least one meter spacing should be given between two plants in each way.

Staking: Growing young yam plants may need staking sometimes. Plants are usually staked on bamboo supports or on trellis.

Harvesting: Yams are ready for harvest within 4 months. Harvested tubers are very tender and brittle and so sufficient care is to be taken at the time of handling.

Figure 10: Purple Yams

Storage: If yams are stored in a cool, dry, hygienic place, it will last for several months. The best storage temperature is 14 to 15 Degree Celsius.

Medicinal Properties of Yams: Not all yams have medicinal properties. Some wild species of Dioscorea are known to have medicinal properties. Their extracts are used for treatment of arthritis.

Amorphophallus: The Elephant Foot Yam

Elephant foot yam is a popular tuber vegetable in tropics and subtropics. It is also known as "white spot giant arum", "stink lily", and "sweet yam". In India, it is known by various names such as Oal, Suran, Zimmikanda/Jimmikand, Kuch. In Sanskrit, it is known as 'Bahukanda' and in Urdu, it is called 'Zamin-kand'. Elephant foot yam is popular in tropical regions such as Africa, South East Asia, New Guinea, Oceania, Madagascar and the Pacific islands as a vegetable crop.

Figure 11: Elephant foot yam plant

Taxonomy: A detailed taxonomic classification of the vegetable is as given below:

Kingdom	Plantae
Clade	Angiosperms
Clade	Monocots
Order	Alismatales
Family	Araceae
Genus	Amorphophallus
Species	A. paeoniifolius
Binomial name	Amorphophallus paeoniifolius
Synonym	Amorphophallus campanulatus

Major Plant Characteristics: The plant consists of single stem and single leaf. It produces a single, large flower and a single large, round-shaped, underground tuber or corm. Plant grows from a modified underground stem which is botanically known as 'corm'. Corm is a modified underground stem which serves as the storage organ for the plants. In case of elephant foot yam, a single stem with a large, umbrella-shaped, multi-lobed, single leaf arises from the underground corm. However the most distinguishing feature of the plant is its large, foul-smelling flower

Origin and Distribution: Elephant foot yams are believed to be originated in the regions comprising of Island Southeast Asia, from there westwards to Thailand and India. It is further distributed westwards to Madagascar, and eastwards to coastal New Guinea and Oceania. It is now commercially grown in India, Sri Lanka, Bangladesh, China, Malaysia, Java, Philippines and Ceylon

Commercial Significance: Elephant foot yam tubers have a long shelf life and a variety of value-added products can be prepared from this tuber vegetable. It has huge production potential and therefore grown as a major cash crop in many tropical and subtropical countries. The plant is grown as an annual herb for raising a commercial crop. The plant is known for its:

- Shade tolerance
- Easy cultivation
- Disease-pest resistance
- Long shelf life of tubers
- High productivity and therefore high profitability
- Variety of value-added products
- Export potential
- Constant demand of tubers in both domestic and global markets

-   Good market price and return on investment

Plant Description: Chromosome number is 28. Elephant foot yam plant usually consists of a single stem and a single leaf. Stem is fleshy and succulent, light green in colour and variegated with creamish white spots. Stem may reach up to 1 to 2 meters height upon maturity. Leaf is highly lobed and is like an umbrella which is about 50 cm in diameter and consists of several oval leaflets. The plant produces a single, large underground corm which is used as a tuber vegetable. Corm is brownish dark in colour and round shaped. Average weight of a corm is usually 3 to 9 Kg. Sometimes the corm or yam tuber may weigh as high as 20-25 Kg

Flower Description: Elephant foot yam plant flowers after its vegetative growth is over and the leaf has completely withered away. It flowers only once in a year and the flower has a life of only 5-days. Flowering season is at the onset of monsoons. First, a large, purple-coloured flower bud emerges from the hidden underground corm. This flower bud slowly develops into a purple inflorescence with both the pistillate (female) and staminate (male) flowers on the same inflorescence. At this stage the inflorescence looks like a large cylindrical mass, top part of which secretes a mucus-like, foul-smelling substance (nectar) and heat that attracts pollinating insects. The middle part of the inflorescence contains staminate flowers and the lower part contains pistillate flowers

Fruit Description: Insect pollination occurs within 2-3 days of flower opening. Female flowers develop into fruiting bodies soon after insect pollination. Fruits are berries and ovoid shaped. They are bright red in colour when ripe

Nutrition in Tubers: A detailed account of nutrition in the tubers of elephant foot yam is as given below:

- 100grams of edible portion of elephant foot yam tubers provides about 116 kcal of energy
- It contains Moisture or Water (70-80g/ 100g), Carbohydrates (18-24g/100g), Protein (2-5g/100g), Fat/Total Lipid (0.2-0.4g/100g), and Crude fiber (0.8-1.68g/100g)
- *Minerals Present:* Phosphorous (34 mg/100 g), Calcium (50-56 mg/100 g), Iron (0.6mg/100g),
- *Vitamins Present:* Vitamin A (434 IU/100 g), Riboflavin (0.06-0.07mg/100g), Thiamine (0.06-0.07mg/100g) and Nicotinic acid (0.7mg/100g)
- Its tender leaves contain 2-3 % protein, 3 % carbohydrates and 4-7 % crude fibre

Nutrition in Flour/Dried Tuber Powder: Researches reveal that dried and powdered tuber flour contains approximately crude protein (1.1%), crude fat (1.1%) and crude fiber (3.5%). Minerals preset in the flour are approximately Phosphorus (1445 mg/kg), Calcium (8536 mg/kg) and Magnesium (1512 mg/kg)

Health Benefits: A detailed account of health benefits of elephant foot yam is as given below:
- Elephant foot yam is a fiber-rich, low fat food and hence can be used as a weight loss diet
- Anti-coagulant, anti-cancerogenic, anti-diabetic, anti-ageing and anti-inflammatory properties
- Lowers cholesterol and boosts immunity
- Increases memory and concentration
- Detoxifying properties
- Presence of omega-3 fatty acids which increase the good cholesterol levels in the blood
- Presence of antioxidant Diosgenin, a molecular hormone which is anti-carcinogenic

Food Uses: Because of the high carbohydrate content, it can be consumed as a staple food after through boiling/steaming; boiled/steamed tubers are mashed and eaten with chutneys. These tubers are suitable for baking and roasting also. Raw tubers are cut into thin slices and deep fried in the hot oil to make chips; these yam chips are a delicacy in many tropical regions of the world. Its tender leaves may be used as a leaf vegetable. Tubers are also used to prepare various vegetable preparations. Elephant foot yam tubers are used in different types of pulses to prepare delicious 'dal' preparations. In certain parts of the world, elephant foot yam tubers are used in pickles and chutneys.

Health Risks and Remedies: Researches show that elephant foot yam tubers contain anti-nutrient factors such as oxalates, tannins, cyanide and phytates in varying levels. The presence of oxalic acid/oxalates in the elephant foot yam may be as high as 1.3% in certain cases. Tannins level may be up to 0.4%, cyanide (35878 ppm) and phytates (0.165%). Thorough cooking is advised because of the presence of 'raphides' (calcium oxalate needles) in elephant foot yam tubers. Calcium oxalate imparts an acrid taste to the tubers and if it is present in the cooked tubers, it gives an itchy sensation in the mouth and throat. Calcium oxalates and the resultant acridity can be removed by boiling the tubers for a long time. Sometimes, boiling tubers with water containing baking soda removes all calcium oxalate needles from the tubers; after boiling, the water is discarded

Medicinal Uses: Elephant-foot yam is believed to have antibacterial, antiprotease, analgesic, and cytotoxic properties. It is traditionally used in Indian Ayurvedic medicines for treating a number of diseases and health disorders. It is widely used as a remedy for breathing disorders such as bronchitis and asthma and stomach disorders such as abdominal pain and dysentery. It is found effective in treating emesis, enlargement of spleen, piles, elephantiasis, and rheumatism also.

Growing Practices for Elephant Foot Yam

A detailed description of growing requirements for elephant foot yam plants is as follows:

Climate: It is very easy to grow elephant foot yam plants. There are many varieties (with flesh colour ranges from white to pink) available for cultivation; the variety/cultivar suitable for local conditions should be chosen for cultivation. Since elephant foot yam is a tropical plant, it needs warm, humid weather with an annual rainfall of 1000-1500mm for its healthy growth. Optimum atmospheric temperature requirement is 25-350C during vegetative growth. During tuber formation, somewhat dry climate is preferred

Soil: Well-drained, well-aerated, deep, fertile, sandy-loam soils that are rich in organic matter is the best for growing elephant foot yam. Water logging and heavy soils should be avoided. Optimum pH is 5.5 to 7

Propagation: Propagation is by the division of mother corms or mother tubers. Offsets or daughter corms (small corms arise from the mother corm) may also be used as propagation materials but they may not produce large tubers. The size of the planting material has a strong effect on the final size of the harvested tubers

Preparing Planting Materials: Mother corms are chosen based on certain parameters:
- Mother corms should be from the previous harvest
- They should be healthy with several potential growth buds/sprouts
- They should be free from all kinds of injuries, bruises, and damages
- They should be free from insect-pest infestations and plant pathogens

Each mother corm is cut into 4-5 pieces of equal size (i.e. setts). Each sett should ideally weigh about 750-1000 g and each should be having a healthy centrally-positioned growth bud. Now all these pieces are smeared with cow dung slurry or wood ash for disinfestation purposes and then allowed to dry in partial shade. Dried planting materials are used for planting in the fields.

Site Preparation: Planting site is prepared b tilling and ploughing twice or thrice during the beginning of the year in the tropics. At the time of land preparation, FYM@25-30tons/ha may be incorporated into the top soil to enhance soil fertility. Pits of size 60x60x45cm at a spacing of 90x90cm are prepared and allowed to weather in the open sun for a few weeks. Pits are filled up with sufficient amount of well-composted FYM (farm yard manure) or cow dung just before planting. The plant may also be grown on flat beds or broad ridges. Broad ridges are suitable for high rainfall areas as it helps in excess water drainage. Higher yields are obtained when crop is grown on raised beds rather than on flat beds.

Figure 12: Tubers of Elephant Foot Yam

Planting: Planting is done during early spring or in the beginning of the year in the tropics (February-March). Pits are half filled with top soil, farmyard manure @ 20-25 kg/pit and other recommended NPK fertilizers just before planting. Planting materials/tubers are placed vertically in pits and filled with rest of the fertilizer-mixed top soil. After compacting the planted tubers, pits are covered with organic/bio mulches like green leaves or paddy straw. After that, a light irrigation is done. Mulching is an important practice in yam-growing as it suppresses weed growth and conserves soil moisture. The planting should be done at 90 x 90 cm spacing for a commercial crop. The best results are obtained when weight of the planting material is about 500g.

If daughter corms/offsets are used, approx.35, 000 corms will be sufficient to plant one hectare area.

Note: Conversion Table as follows:

1 hectare/ha	10000 sq.m (square meter)
1 acre	Approx. 4000 sq.m

Fertilizers: Recommended fertilizer rate for a commercial crop is 80:60:80 NPK/ha. Fertilizer should be applied in split doses. Full doses of Phosphorus (P) and half doses of Nitrogen (N) and Potash (K) are applied at the time of planting. Remaining half doses of N and K are applied 60-70 days after emergence of shoot/stem. After the second fertilizer application, one earthing up is done

Irrigation: In tropics, elephant foot yam is raised as a rainfed crop. However, when rainfall is deficient, frequent light irrigations are given, especially during the initial vegetative growth stages. Towards the tuber-formation stage, irrigation is required very less

Weed Management and After-Care: Weed control is an important cultural operation in yam-growing. Hand weeding is normally practiced. Earthing up and other cultural practices also control weed growth up to a great extent. Mulching is an important operation that controls weed growth effectively.

Time to time earthing up and hoeing is done. If a plant produces more than one shoot/stem, the extra shoot should be removed

Disease-Pest Management: Elephant foot yam is susceptible to fungal diseases such as collar rot/foot rot and viral diseases such as mosaic. The plant is susceptible to a number of insect-pests such as aphids, mites, thrips, mealy bugs etc. Observing proper cultural practices may reduce disease-pest incidences in the fields up to a great extent

Disease Management: A detailed account of various disease management practices is given below:

Collar Rot: Collar rot/foot rot is a fungal disease caused by Sclerotium rolfsii and Rhizoctonia solanii. It is prevalent in high rainfall and high humid areas where water drainage is a major issue. Young plants of 2-3 months old are highly susceptible to this disease. Collar region of the plants is attacked by the fungal pathogens and soon water soaked lesions appear on stems and eventually the whole plant turns yellow and rots; finally the plant collapses and dies. Best control measures include soil treatment with a recommended fungicide, providing good water drainage in the growing fields, crop rotation and treatment of the plants with biocontrol agents like *Trichoderma harzianum* @2.5kg/ha. *Trichoderma harzianum* should be applied with large quantities of farm yard manure (FYM)/compost; the entire recommended quantity should be mixed with 40-50 tons of FYM before application

Mosaic Disease: Mosaic is a viral disease that is primarily spread by planting the infested tubers and it is further spread across the growing fields by vector flies such as aphids. Mosaic mottling of leaves and abnormal leaf development are the major symptoms. Affected plants do not grow properly and corm production is also poor. The best control measures include removal of diseased plants from the fields and ensuring using only disease-free planting materials. Control of aphids helps in preventing the spread of the disease

Insect Pest Management: Aphids can be effectively controlled by introducing their natural enemies such as ladybugs in the fields. Caterpillars, Mites, Mealy bugs and Thrips can be controlled by spraying neem oil-based emulsions, diluted soap solution or chilli-garlic extract, or similar bio insecticidal formulations. Spraying with a recommended insecticide is advised in case of severe insect-pest infestations

Harvesting and Yield: The crop is ready for harvest within 7-8 months after planting. Towards the maturity, the leaf turns yellow and starts withering. When the leaf has completely dried off, the tuber can be harvested by using a spade to dig out the full-grown tuber very carefully. Yield per hectare is about 20 – 40 tons/ha. The harvested tubers may be stored in well-aerated, cool, dry, hygiene shade places for several months.

Tannia or Xanthosoma spp.

Tannia or Xanthosoma is another edible aroid crop. It is believed to be originated in South America. Two main species of Xanthosoma are X. sagittifolium and X. violaceum. X. sagittifolium is mainly cultivated as a food crop. It can be grown as a biennial or perennial crop. Its growing practices are more or less similar to that of taro plant.

Propagation of the plant is through setts or cormels. Soil requirements and water requirements are same as that of a taro plant. Filed is prepared by continuous ploughing followed by furrow making.

Spacing: Xanthosoma is planted on ridges or furrows at a spacing of 60 cm between plants in the row and 1 meter between rows. When planted on flat land, a spacing of 1 meter x 1 meter may be used.

Harvesting: Harvesting maturity is indicated by yellowing of the leaves. Plants come to maturity about 10–12 months after planting. Complete uprooting of the plant is not necessary for harvesting mature corms. While harvesting the mature corms, small corms and cormels may be left undisturbed for

future harvesting. A tannia plant may be harvested
continuously up to 5-6 years.

Pest and disease management practices are similar to that of a
taro plant.

Swamp Taro or Cyrtosperma spp.

Swamp taro or Cyrtosperma merkusii is another important edible aroid crop which is believed to be originated in Indo-Malayan region. It is mainly grown as a seasonal crop for its edible corms. Leaves are not used for food purposes. Growing practices are more or less similar to that of tannia crop.

Lasia spinosa or Kohila.

Lasia is a lesser known edible aroid crop which is a native to Asia and New Guinea. Both its corms and young leaves are used as food in many Asian countries. In Srilanka, lasia is also known by the name kohila.

Figure 13: Kohila vegetable